THE IRON BRI

Geoff Alton

The Earls High School,
Hales Owen

CONTENTS

Introduction	1
How to use this book	2
The Bridge	4
The town of Ironbridge	19
Conclusions and further research	30
Further reading and acknowledgements	32

INTRODUCTION

The Iron Bridge which crosses the River Severn half a mile to the east of Coalbrookdale, Shropshire, is today one of the country's leading tourist attractions. Visitors from all over Britain and the world come to see what the historian Barrie Trinder has called 'the most widely recognised symbol of the Industrial Revolution'.

But making a visit to the Iron Bridge is not merely a twentieth century experience. Ever since traffic first crossed over the bridge, on New Year's Day 1781, people have been coming to look at the world's first cast-iron bridge. Artists have brought their sketch books, brushes and paints, writers their notebooks, and engineers their theodolites and tape measures.

At the time the bridge was not only a world first but a wonder of the age. Stone and timber had for centuries been the only materials used for building bridges (and most other things as well). Never before had anyone thought about, let alone built, a bridge of cast iron. The building of the bridge marked a turning point, not only in civil engineering but in the future of industrialisation. Just as coal had replaced charcoal and become the new fuel of the age, and steam had replaced water as the source of power, so too iron replaced timber and stone as a building material. Perhaps more than anything else the building of the Iron Bridge showed the world the potential of iron as a constructional material. In the following decades new factories, and the machinery inside them, would be made of iron. So too would ships, canal boats, window frames and even chimney pots.

It is therefore perhaps fitting that the Iron Bridge should be linked with the famous iron-making families of the Darbys and the Wilkinsons, both of whom did so much to revolutionise the production of iron in the Severn Gorge and throughout Britain.

If the only effect of the Iron Bridge had been to popularise the use of iron, that alone would have made it important in the history of Britain's industrialisation. However, it is important to realise that the bridge was constructed for a practical purpose – to link the communities on both sides of the River Severn. It also marked the beginning of a new town – Ironbridge – which itself generated new trade, industry and jobs for the people of the Severn Gorge.

HOW TO USE THIS BOOK

To complete the book you need no more than a pen or pencil though you should take a camera if you wish to record the area in greater detail. A number of postcards of the bridge and surrounding area can be purchased from the Museum shop along with additional Museum books on the area. When completed, this book forms a personal record of your visit to the Iron Bridge.

The main aim of the book is to involve the reader in an investigation of the Iron Bridge and the surrounding area. Your exploration of the area covered by this book should be treated as a detective exercise. Look for clues which will enable you to learn something about the history of the site. Also ask questions and pose your own theories about the physical evidence you are studying. Although the general idea is for you to investigate, interpret and evaluate the *physical evidence* in front of you, nevertheless a number of different types of documentary sources have been included in the book to enable you to reconstruct a detailed picture of the area as it appeared in the eighteenth and nineteenth centuries.

Map of the area to be explored

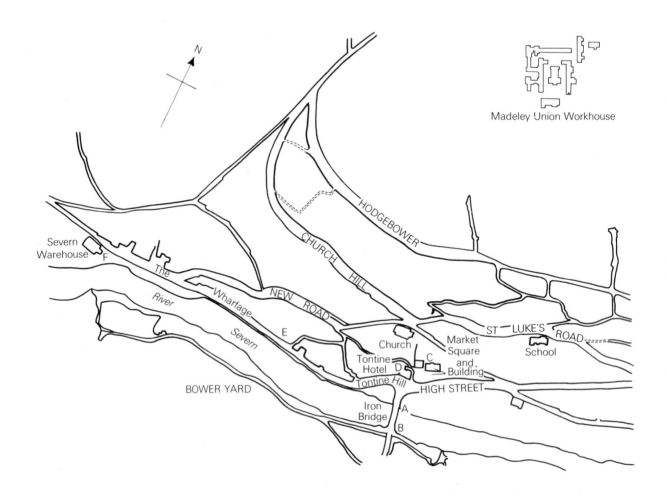

A = The Iron Bridge
B = Toll House
C = Market Building
D = Tontine Hotel
E = The Wharfage
F = The Severn Warehouse

Start of exploration

The start of your exploration is the Iron Bridge marked by the letter A on the plan opposite. Your first task is to familiarise yourself with the surrounding area and, in particular, to identify the features you will be studying in some detail.

1. Take up a position on the Iron Bridge which enables you to look onto the town of Ironbridge. Study the plan opposite and aerial photograph below. The letters on the plan are the main buildings/features you will study on this walk. Use the plan to identify all of the features labelled around the edge of the aerial photograph. Draw arrows to these features.

2. Briefly describe the area all around you.

From the side of the Iron Bridge there is a lovely view with the Toll House which now you can buy things from. There is another bridge which is similar to the Iron Bridge except built later

Wharfage   Severn Warehouse   Tontine Hill   Tontine Hotel   St Luke's Church

Church Hill

Market Square and Building

High Street

Bower Yard   Toll House   Iron Bridge   River Severn

THE BRIDGE

A THE IRON BRIDGE

First impressions

A	How were you impressed?
B	When was it built?
C	Who built it?
D	Why is it built here?
E	How much did it cost?

It is recommended that you spend about ten minutes exploring the bridge before working through the exercises that follow. Your preliminary investigation should take you both over and under the bridge. During this time you should begin to ask your own questions about the bridge and look for clues that might help to answer these questions.

1. Make a list of five questions you would like to ask about the bridge at this stage of your investigation and put them in the box above.

2. How many of your five questions could be answered by studying the bridge itself and how many require the use of documentary sources? List them

....
....
....
....

> But of the Iron Bridge over the Severn, which we crossed, and where we stopped for half an hour, what shall I say? That it must be the admiration, as it is one of the wonders of the world.
> *John Byng, later Viscount Torrington, 1784*
>
> ...we noted some of the regions most remarkable objects, and among them of course the Iron Bridge which was a general curiosity to me.
> *Charles Dibdin, c.1801*

During the late eighteenth and early nineteenth centuries a number of visitors to the bridge left behind their thoughts and feelings about their visit. It is interesting to note that although the bridge continued to be a tourist attraction throughout much of the nineteenth century not all visitors shared the enthusiasm of John Byng and Charles Dibdin. A Swedish visitor in 1802 claimed that,

> ...this Bridge is less remarkable with regard to its appearance and size since several bridges have been built in England which are larger and more imposing.

3. Why do you think that John Byng and Charles Dibdin saw the Iron Bridge with astonishment and wonder whereas some other visitors were much less impressed.

   ....
   ....
   ....
   ....
   ....
   ....

4. What are your thoughts/feelings about the bridge?

   ....
   ....
   ....
   ....
   ....
   ....

Why build a bridge in the Severn Gorge?

Before the Iron Bridge was opened the nearest bridges to Coalbrookdale were at Buildwas (three km upstream) and Bridgnorth (14 km downstream). Although the Severn Gorge had been a busy industrial area for nearly two hundred years before the eighteenth century, the shortage of bridges does not appear to have been a cause of much concern among the then tiny communities of the region.

Not until the mid-eighteenth century was there talk of a bridge. By this time the whole of the Severn Gorge was a hive of activity. The pace of industrial growth had increased greatly.

This was the period that saw the rapid expansion of the iron industry in the region, the products of which were, by the late eighteenth century, in great demand throughout Britain. At the same time there was a rapid expansion of glass making, lead smelting, brick and pottery making in the area.

With this growth in trade and industry came the growth of the many riverside communities. More than ever before there was a need for a bridge to link the communities that were developing on both sides of the river.

There were, of course, ways of crossing the river other than by a bridge. At least four ferries operated in the Gorge and there were numerous privately owned boats of all shapes and sizes, as was to be expected in riverside settlements. The problem with these methods of transport was, however, obvious to

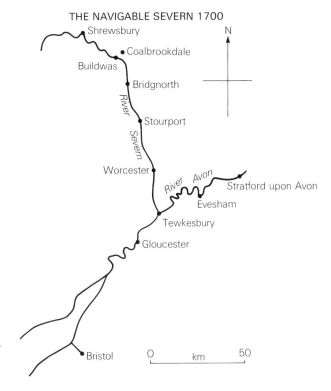

THE NAVIGABLE SEVERN 1700

all whom had used them. The Severn is a river that can rise rapidly after a few days rain and flows with a swift current through the narrow valley of the Severn Gorge. In times of flood, crossing the river by ferry was extremely dangerous, as the following account written in 1704 shows:

> The river was in full flood and the wind did chill my bones. I paid the ferryman and he pulled away from the shore. The journey normally was no more than a ten minute episode but we were pushed hither and thither by the wind tossed waves. I clung to the sides as the little vessel was swept down stream ... after a full three quarters of the hour we reached the shore.
>
> *Shadrax Fox, 1704*

1. What time of the year do you think this event happened and why?

   ....
   ....

Shadrax Fox completed his journey across the river as did hundreds of other people every year. Not so fortunate were the people using the Coalport ferry (a few kilometres downstream from Coalbrookdale) on the night of 22 October 1799. This ferry overturned and 27 people drowned.

When you know about the dangers that were, and still are, involved in crossing this part of the River Severn by small boat or ferry, it is not surprising that a bridge was built in the Gorge. What *is* surprising is that the idea for a bridge did not come sooner and that when it did the bridge was built out of iron, and not the traditional materials of stone, brick and timber.

Background to the project

In 1775 a group of people who were interested in building a bridge across the Severn Gorge met to decide on how to proceed with the project. Among the men first involved with the new bridge were the fourteen listed below, all of whom agreed to help pay a share of the cost of building.

Ironbridge Sharelist 1777: 64 shares of £50 per share

Revd Harris	10	John Morris	2
Abraham Darby	15	Charles Guest	2
John Wilkinson	12	Roger Kynnaston	1
Leonard Jennings	10	John Hartshorne	1
Samuel Darby	4	Sergeant Roden	1
Edward Blakeway	2	John Thursfield	1
Farnolls Pritchard	2	John Nicholson	1

After many meetings it was agreed that Thomas Farnolls Pritchard, a Shrewsbury architect, would design the bridge and Abraham Darby III of the Coalbrookdale Ironworks (grandson of the first Abraham Darby) would build it.

On 5 February 1776 the shareholders sent a petition to Parliament asking for permission to build a bridge across the Severn.

In March of the same year the petition became an Act of Parliament, the first page of which is shown opposite.

1. Why, according to the Act, was there a need for a bridge across the Severn?

   ....
   ....
   ....

2. Why do you think the Act mentions that the owners of the existing ferry were willing to allow the bridge to be built? (Look at the last eight lines of the document.)

   ....
   ....
   ....
   ....
   ....

ANNO DECIMO SEXTO

Georgii III. Regis.

C A P. XVII.

An Act for building a Bridge across the River *Severn* from *Benthall*, in the County of *Salop*, to the opposite Shore at *Madeley Wood*, in the said County; and for making proper Avenues or Roads to and from the same.

WHEREAS a very considerable Traffick is carried on at Coalbrook Dale, Madeley Wood, Benthall, and Broseley, in the County of Salop, and the Places adjacent, in Iron, Lime, Potters Clay, and Coals, and the Persons carrying on the same are frequently put to great Inconveniencies, Delays, and Obstructions, by reason of the Insufficiency of the present Ferry over the River Severn from Benthall to Madeley Wood, commonly called Benthall Ferry, particularly in the Winter Season, in which Time it is frequently dangerous, and sometimes impassable: And whereas the Reverend Edward Harries and Abraham Darby are Owners of the said Ferry, who, with the several Persons herein-after named, are willing and desirous,

Preamble.

5 L 2

Thomas Farnolls Pritchard

3. Suggest three reasons why Abraham Darby bought the largest number of shares in the project.

   (i) ....
   ....
   ....

   (ii) ....
   ....
   ....

   (iii) ....
   ....
   ....

The siting of the bridge

A view of the River Severn, near Coalbrookdale, where the iron bridge is to be built, *by William Williams, c. 1776*

It would appear that there were only two possible sites for a bridge, both of which were ferry crossings and far from ideal.

The problem, as can still be seen today, was caused by the steep sides of the heavily wooded valley. The Severn Gorge is a narrow valley with the river lying about 40 metres below the surrounding area.

It was not just a question of finding the narrowest and shallowest part of the Severn (both of which make the task of bridge builder that much easier). The bridge also had to link up with existing road networks on either side of the river, or at least make it possible for new stretches of road to be built. The scale of the problem can be seen by comparing the time that it took to erect the bridge with the building of new road links. All the main parts of the bridge were erected in three months. It took a further two years to complete the approach roads and link-roads.

Look carefully at the picture above which shows the site eventually chosen for the Iron Bridge. The Tontine Hotel was built in the area shown in the top right-hand corner of the picture. Note the absence of any buildings in the bottom half of the picture, where the bridge was to be built.

1. Why would the financiers and builders of the bridge have welcomed this lack of buildings around the chosen site?

   ....
   ....
   ....
   ....

2. What evidence is there in the picture to support the statement that this part of the Severn Gorge was a 'hive' of industrial activity even before the bridge was built?

   ....
   ....
   ....
   ....
   ....
   ....
   ....
   ....

7

Construction and detail

Engineers have different names for the way bridges are constructed.

1. Look carefully at the various types of bridges shown in the pictures and then answer the question, 'What type of bridge is the Iron Bridge?'

   ....

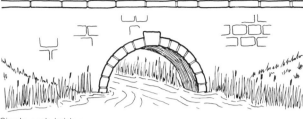
Single arch bridge

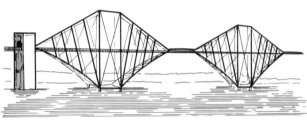
Cantilever bridge

Suspension bridge

Multi-arch bridge

Movable bridge

2. Suggest two possible reasons why the Iron Bridge was not built to one of the other designs shown.

   (i) ....
   ....

   (ii) ....
   ....

The photograph below shows the main structural features of the Iron Bridge, two of which have been correctly identified for you. Find and examine the other features labelled around the edges of the photograph.

3. Draw arrows to these features as you locate them.

4. Complete the following tables:

INFORMATION REQUIRED	ANSWER	INFORMATION REQUIRED	ANSWER
Number of main arches		Number of uprights	
Number of radials		Number of small ribs	
Number of ogees		Number of cross stays	

5. What prevents the main arches from moving sideways?

   ....
   ....

6. How are the base plates secured to the foundation stones?

   ....
   ....

7. How are the railings fixed to the bridge?

   ....
   ....

8. Look carefully at the bottom of the main arches on the north abutments (ribs on the north side of the river). Notice the grooves made in the cast iron. These were gradually worn into the ribs during the nineteenth century. What do you think caused this?

   ....
   ....
   ....
   ....

An Italian visitor to the bridge in 1787 wrote the following in his journal:

> The Iron Bridge ... has a beautiful appearance on both sides – has proved very strong – but it is not built with Mathematical truth as the inner arch or rib in the main support of the two parallel ones, being no addition of its strength but much weight.
> *Carlo Castone, 1787*

It is a fact that the weight of iron in the bridge is much greater than was used in later iron bridges of similar size.

9. Make a list of any other parts of the bridge's structure which you think do not add to its strength.

   ....
   ....
   ....

10. Suggest reasons why Abraham Darby decided to use more iron than was necessary

    ....
    ....
    ....

If you look carefully at the way in which the different parts of the bridge have been put together you will notice that not a single screw, nail or rivet has been used. Each part is slotted into or through each other. Any movement in the joints is taken up by using wedges made of cast iron or jointed by fitting into dovetail boxes. In short, the world's first iron bridge was put together using the traditional skills of the carpenter and stone mason.

Today it would be unthinkable to use carpentry joints when constructing a bridge made of iron but, at the time it seemed logical to use building methods that had been in use for centuries, once the decision had been made to built the bridge out of iron. From the available evidence it appears that the bridge trustees were, for a time, undecided about what materials to use, as can be seen from the advertisement on page 10. This appeared in a number of Midlands' newspapers in May 1776.

11. Is this advertisement evidence that the Bridge Trustees did not decide on a bridge of iron until after May 1776? Explain.

....
....
....
....
....

Read the extract below from the Bridge Trustees' minute book.

Clerk of the General Meetings.
N.B. *Serjeants will attend from Twelve to Six in the Afternoon of the preceding Day, to provide those with Quarters who may come in on that Day.*

A BRIDGE to be Built.

ANY Perfon willing to undertake to Build a BRIDGE, of one Arch over the Severn, from Benthall Rail to the oppofite Shore in Madeley Wood, in the County of Salop; of Stone, Brick, or Timber, the Arch One Hundred and Twenty Feet in the Span, the Superftructure Eighteen Feet Clear, the Centre of the Arch, Thirty Five Feet from Low Water; are defired to fend Propofals to Thomas Addenbrooke, at Coalbrookdale, before the 20th of June next, and to Attend a Meeting of the Proprietors, at John Nicholfon's, at Coalbrookdale, on Friday the 28th of the fame Month, at Eleven o'Clock in the Forenoon.
Coalbrookdale, 15th May, 1776.

> 18 October 1776.... Abraham Darby agreed to erect an Iron Bridge of one Arch one hundred and twenty feet span and the superstructure not less than eighteen feet wide ... to be completely finished with roads ... to and from the same as described in the Act of Parliament, on or before the twenty fifth day of December 1778.
> *Minute Book of the Proprietors of the Iron Bridge*

---

12. What changes had been made to the construction of the bridge between May and October 1776?

....
....

13. Is this source evidence that no one responded to the advertisement?

....
....
....

14. Did Abraham Darby fulfil the terms of the agreement? Explain.

....
....
....
....

It has been suggested by some historians that Abraham Darby took part in the bridge project because he wanted his iron company to build a bridge of iron across the Severn. By doing this the bridge would not only help the development of the riverside communities, but, more importantly, act as a good advertisement for the Coalbrookdale Ironworks of which Darby was a partner.

15. If the above is correct how would the building of an iron bridge have acted as an advertisement for the Coalbrookdale iron company?

....
....
....

16. In the box below sketch at least two of the different types of joint used in the construction of the Iron Bridge. Use the terms on page 8 to label correctly the precise parts of the bridge held together by the joints. For example, the joint used to secure the main rib to the base plates.

17. Why do you think that it seemed 'logical' to use carpentry joints at the time the bridge was constructed?

    ....
    ....

18. Why don't bridge designers today use these joints?

    ....
    ....

Making the parts for the bridge

Having solved the problem of how to construct an iron bridge across the Severn, the next task facing Abraham Darby was how to make the various parts. Unfortunately there is not enough evidence to enable the historian to say for certain where the parts for the bridge were cast or the problems, if any, they had with the castings. The evidence that *is* available suggests two possible theories as to where the parts were cast.

A   The parts were made at the upper furnace site at the Coalbrookdale ironworks, about one kilometre away. (See map opposite.)

B   The parts were made on the banks of the River Severn close to where the bridge was to be built.

Points To Consider

- Each of the main ribs weigh over five tonnes (at least 8 times the weight of the average family car). In all approximately 384 tonnes of iron was used.
- Every single piece had to be cast. This involved pouring molten iron into wooden moulds or, as shown in the picture below, casting direct into moulds made in sand.
- The molten iron could come direct from a large blast furnace, like the one still to be seen at Coalbrookdale, or from smaller portable (or cupola) furnaces like the one shown opposite.
- Cast iron is very fragile and it is impossible to repair once broken.

The inside of a smelting house at Broseley, by G. Robertson

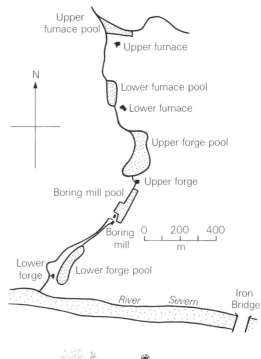

A cupola furnace

1. Take into consideration the information above and then state which of the two theories you support and why.

   ....
   ....
   ....
   ....
   ....
   ....
   ....
   ....

11

Building the bridge

One of the frequently asked questions about the Iron Bridge is how was it actually put up across the Severn? Although there are only brief references to the erection of the various parts, nevertheless we can gain an insight into the possible methods used by studying contemporary diagrams of lifting gear and methods used to transport heavy goods.

The picture of the inside of a smelting house on page 11, for example, shows a late eighteenth century crane used for lifting heavy iron castings. The pictures opposite show other ways of lifting and moving heavy objects around during the same period.

So much for the 'machines' and other equipment, but what about the manpower? While much is known about the men who financed the project very little is known about many of those who contributed most to the actual building of the Bridge. On this subject the visitor has to do what the historians Barrie Trinder and Neil Cossons did in the extract below – use his or her background knowledge of the period and something of their imagination.

> Stonemasons toiled a whole summer on the high abutments; pattern makers created wooden patterns for huge castings with sufficient accuracy for ribs to be threaded through uprights and fitted into dovetail joints; moulders prepared moulds of sizes and shapes to which they were totally unaccustomed. Never before had molten iron flowed into shapes like those of the ten half ribs of the bridge and only steam-engine cylinders among the usual products of the foundry were comparable in weight. Thomas Sutton and his crew manoeuvered their barge between the abutments of the bridge, probably carrying the ribs on board. Teams of labourers created a scaffolding of dizzy height, and doubtless every man on the payroll, together with teams of heavy horses strained on ropes to lift the ribs into position . . . others . . . hacked through rocks on the side of the Gorge to link the bridge with the Madely turnpike.
> *Extract from* The Iron Bridge *by B. Trinder and N. Cossons, 1979*

1. Below is a list of some of the main stages in which the bridge was constructed. Put them in the order in which you think the bridge was erected starting with F 'Clearing the ground around the site'.

4. Apart from iron what other materials have been used in the bridge's construction?

....

....

A.	Casting the iron ribs and other parts.	1. = F
B.	Erecting scaffolding.	2. =
C.	Fixing the base plates to the floor.	3. =
D.	Fixing the two sets of uprights and cross stays.	4. =
E.	Building the stone abutments.	5. =
F.	Clearing the ground around the site.	6. =
G.	Lifting and putting into place the main rib section.	7. =
H.	Fixing the ogees and radials.	8. =
I.	Fixing shorter ribs to the main ribs.	9. =
K.	Building the road across the bridge.	10. =

2. What do you think was the most difficult stage to complete and why?

....
....
....
....
....
....

3. How do you think this difficulty was overcome?

....
....
....
....
....

5. There are two inscriptions to be found on the bridge stating the year in which the bridge was erected. Where exactly are these inscriptions to be found?

(i) ....
....
....

(ii) ....
....

6. One of the inscriptions is in Roman numerals. Work out the value of the following numerals:

M =   ; D =   ; C =   ; L =   ;
X =   ; IX =   .

13

7. Use this page for a sketch of the motif found on the railings of the bridge, and additional photographs and/or postcards of the bridge.

8. Complete the partly drawn sketch of the Iron Bridge on the next page, filling in the required labels and, if possible, measurements.

SKETCH OF THE IRON BRIDGE LOOKING DOWNSTREAM

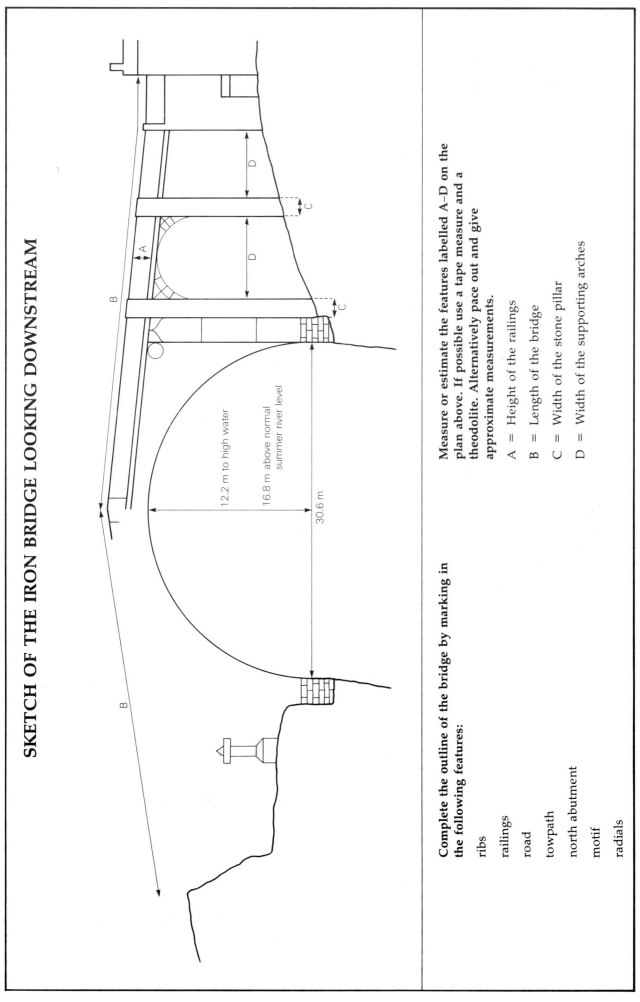

Complete the outline of the bridge by marking in the following features:

ribs

railings

road

towpath

north abutment

motif

radials

Measure or estimate the features labelled A–D on the plan above. If possible use a tape measure and a theodolite. Alternatively pace out and give approximate measurements.

A = Height of the railings

B = Length of the bridge

C = Width of the stone pillar

D = Width of the supporting arches

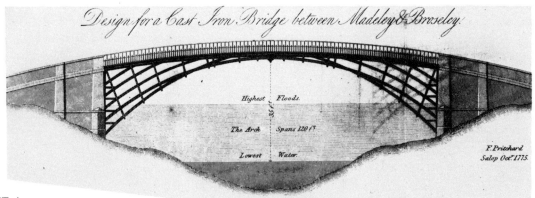

SOURCE A
Pritchard's original design, 1775

SOURCE B
Engraving of Rooker's painting, 1782

SOURCE C
Museum postcard or personal photograph 19— to be put here

The bridge through time

Take up a position downstream (Bridgnorth side of the bridge) which enables you to see the bridge from the same direction as Rooker was in when he painted Source B.

1. Look carefully at Sources A and B. In what ways are the two pictures of the Iron Bridge itself:

    (i) similar ....
    ....
    ....
    ....

    (ii) different ....
    ....
    ....
    ....

Source A is Pritchard's original design for an iron bridge across the Severn, drawn for the Bridge Trustees in 1775.

Source B is the bridge that Abraham Darby actually built across the Severn in 1779.

2. Why do you think that Abraham Darby decided to modify Pritchard's original design? Suggest at least three reasons.

    (i) ....
    ....

    (ii) ....
    ....

    (iii) ....
    ....

    (iv) ....
    ....

Source B shows a number of sailing boats (known as Severn Trows) in the background of the picture and one in the foreground.

3. Does this painting help to explain why the Iron Bridge was designed with only one main arch across the river?

    ....
    ....
    ....
    ....

Look again at Source B and the bridge as it appears today. Note the changes that have been made to the stone abutments shown in Source B. The massive stone abutments, to the right hand side of the picture were in fact replaced by two small wooden arches as early as 1802–4. These in turn were replaced by the present arches in 1821. Quite clearly these were major structural changes to the bridge.

4. Why do you think the Bridge Trustees felt it was necessary to make these changes? Was it (a) for aesthetic reasons (appearance); (b) because the stone abutments began to subside or (c) other reasons?

    ....
    ....
    ....
    ....

5. Apart from the changes mentioned already what other changes appear to have been made since the time of the Rooker painting (Source B)?

    ....
    ....
    ....
    ....

6. Choose any one of the above changes and provide a theory as to why the change took place.

    ....
    ....
    ....
    ....

7. Which other types of historical sources might provide evidence to support or refute your theory? Explain how they might do this.

    ....
    ....
    ....
    ....
    ....
    ....
    ....

B THE TOLL HOUSE

The Toll House (now a Tourist Information Centre) stands on the south side of the Iron Bridge and was built to house the toll-keeper and his family. He was responsible for collecting money, or toll as it was called, from anyone who wanted to cross over the bridge.

1. Apart from the house and the toll board on the outside wall, what other evidence is there that this was once a toll-bridge?

   ....
   ....

2. Examine the toll board, a postcard of which can be bought from inside the toll house. How much would it have cost for a stage coach pulled by four horses and carrying six passengers to have crossed the bridge?

   ....

3. Do you think that this toll board was in daily use up until the day that the bridge ceased to be a toll bridge in 1950? Explain your reasons.

   ....
   ....
   ....
   ....

4. What, if anything, can we learn about the bridge owners' attitudes from the last four lines on the toll board?

   ....
   ....
   ....
   ....

5. Why do you think the Toll House was built with its foundations where they are? Why not on the bridge itself?

   ....
   ....
   ....
   ....

6. Suggest two reasons why the Toll House was built on the south end of the bridge and not the north.

   (i) ....
   ....

   (ii) ....
   ....

Put a postcard or a photograph of the toll board or a view of Ironbridge town in the box below.

THE TOWN OF IRONBRIDGE

Although there was a settlement in this area before the bridge was built in 1779 it was not until after this date that the community became known as Ironbridge or developed into a prosperous market town. (It is important to note that the village of Coalbrookdale and the town of Ironbridge are, and always have been, two separate communities, a fact not always realised by late eighteenth and early nineteenth century visitors.)

The bridge, as we have seen, was built as a direct result of growing local demand at a time of increased trade and industrial activity in and around the Severn Gorge. The effects of the bridge on the surrounding communities were considerable and made more so by the new and improved local road system which was built either during, or shortly after, the construction of the bridge.

With improved communications, trade between the various riverside communities expanded as did trade between the whole of the Severn Gorge and the rest of the country. By the late eighteenth century the area was not only famous for its Iron Bridge but also for its products, such as iron, pottery and coal.

This sketch, done by Joseph Farington in 1789, shows the settlement of Ironbridge in its early stages soon after the construction of the bridge.

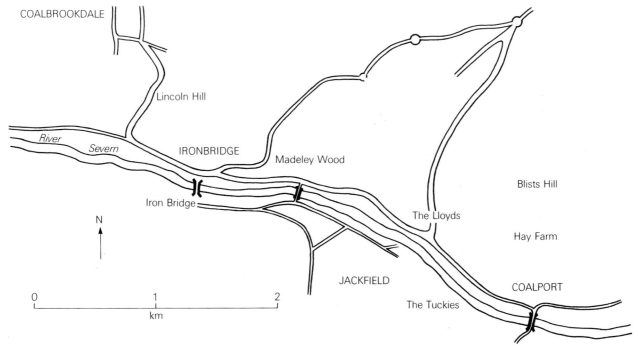

ACADEMIES AND SCHOOLS
(1)   run by Robert Fell

ATTORNEYS
(1)   Matthew Coart

AUCTIONEERS
(1)   Sam. Walter

BAKERS AND CONFECTIONERS
(2)   Edward Jones
      George Milner

BLACKSMITHS
(3)   John Edwards
      John Stormont
      Richard Walton

BOOKSELLER, STATIONER AND PRINTER
(1)   William Smith

BOOT AND SHOEMAKERS
(4)   Thomas Dudd
      Edward Franklin
      William Howard
      James Woodruff

BRAZIERS AND TIN PLATE WORKERS
(2)   Charles Bartlam
      John Fenton

BRICK AND TILE MAKERS
(2)   Roger Cock
      Edward Edwards

BRICKLAYERS
(4)   John Nevet
      Samuel Nevet
      Thomas Nevet
      Thomas Thompson

BUTCHERS
(2)   Thomas Barrett
      Jeremiah Jones

CABINET MAKERS
(1)   Edward Edwards

COOPERS
(3)   Richard Havnes
      Thomas Rogers
      William Williams

CURRIERS
(3)   Henry Brown
      Roger Cock
      Richard Cotton

GROCERS AND TEA DEALERS
(6)   Leonard Crowder
      James Glazebrooke
      Mary Hauley
      Betsey Milner
      Henry Smith
      William Smith

HOP AND SEED DEALERS
(1)   Jas. Parker

INNS
Tontine (and posting)

IRONMONGERS
(2)   Richard Davies
      John Haywood

JOINERS AND CARPENTERS
(2)   William Jenks
      John Lloyd

LINEN AND WOOLLEN DRAPERS
(4)   Henry Charlton
      Edward Edwards
      James Glazebrooke
      William Weare

CARRIERS
To LONDON, Crowley Hicklin & Co. from Ironbridge every week.
To SHREWSBURY, Richard Parry, from the Swan Inn, Ironbridge... every Tuesday and Friday.

MALSTERS
(8)   William Anstice
      James Blanthorn
      Richard Boycott
      George Chune
      Samuel Davies
      James Parker
      Samuel Smith
      Francis and John Yates

NAIL MAKERS
(3)   Richard Andrews
      Richard Armstrong
      Benjamin Jones

PAINTERS, PLUMBERS AND GLAZIERS
(2)   Leonard Crowder
      William Jenks

SADDLERS
(2)   Bejamin Davies
      Samuel Grosvenor

SALT DEALERS
(3)   George Goodwin
      Samuel Poole
      Francis Yates

SHOPKEEPERS AND DEALERS IN SUNDRIES
(1)   Richard Armstrong

STRAW HAT MAKERS
(3)   Jabez Aston
      Elizabeth Delves
      Seth Raby

SURGEONS
(2)   Richard Proctor
      Rowland & Son

TAILORS
(3)   Jas Blanthorn
      Edward Blodwell
      Thomas Delves

TALLOW CHANDLERS
(1)   Benjamin Goodwin

TAVERNS AND PUBLIC HOUSES
Coopers' Arms
Royal Oak
Three Tuns
White Hart

TIMBER MERCHANTS
(2)   George Chane & Sons
      Stephen Davies

WATCH AND CLOCK MAKERS
(2)   James Burrows
      William Liseter

COACHES
To BIRMINGHAM, the Emerald (from Shrewsbury) calls at the White Hart, Ironbridge, every morning at ten.
To CHELTENHAM, the Old Worcester (from Shrewsbury) calls at the Tontine Inn Ironbridge, every Monday, Wednesday and Friday mornings at eight... goes through to Worcester.

An impression of how the town of Ironbridge developed after 1779 can be gained from the local trade directories. These were the equivalent of today's Yellow Pages. In the nineteenth century trade directories were compiled by private firms for the benefit of local businessmen and traders, and also for visitors to the places covered by the directories. On page 20 is a list of the trades and occupations listed in *Pigot's Directory* for the town of Ironbridge in the year 1828.

1. From the Directory it would appear that a small number of people were involved in more than one occupation. List their names and occupations.

   ....
   ....
   ....
   ....

2. Suggest a reason why very few businesses were run by women.

   ....

3. Which of the occupations listed no longer exist today?

   ....
   ....

4. Using only the Directory as your source of information and evidence make three statements about Ironbridge as it appeared in 1828.

   (i) ....
   ....

   (ii) ....
   ....

   (iii) ....
   ....

The general impression given by the Directory is of a prosperous and developed town, an impression supported by Charles Hulbert's description of Ironbridge written a few years later.

> Here we may say is the mercantile part of the town of Madeley, [name of the parish] and here is the focus of professional and commercial pursuits. The weekly Market, the Post Office, the Printing Office, principal Inns, Drapery Grocery and Ironmongery, Watch Making, Cabinet Making, Timber and Boat Building establishments, the Subscription Library, Subscription Dispensary, Branch Bank, Subscription Baths, Gentlemen of the Legal and Medical Professions, Ladies Boarding School etc., etc.
>
> *Charles Hulbert, 1837*

5. Underline the occupations/facilities mentioned in Hulbert's account but not listed in the Directory.

6. Suggest three reasons why these were not listed in the Directory.

   (i) ....
   ....

   (ii) ....
   ....

   (iii) ....
   ....

There is little doubt that Ironbridge was once an important and prosperous market town and centre of communications for the numerous communities in and around the Gorge. The reliability of the general impression, if not the finer detail, created by both the Directory and Hulbert, can easily be cross checked by a brief walk around the streets of Ironbridge today. Although many of the former trades have disappeared, much of eighteenth and early nineteenth century Ironbridge still remains, especially the houses and public buildings associated with a

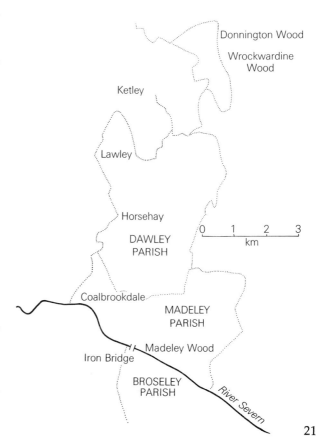

prosperous and thriving town. For example, St Luke's Church, perched on a most unlikely site high up on the slopes of the Gorge, built at a cost of £3232 in 1830: a clear sign of the town's prosperity.

Further evidence of the town's prosperity can be seen in the most fashionable residential area of the town behind the Market Square along Church Hill and the hill-top behind the church. Here the larger houses of Ironbridge's wealthier inhabitants were built, and in a wide variety of architectural styles.

If time permits you should make a tour of upper Ironbridge to explore the many fascinating features of what is an unusual site on which to build a settlement. Only then will you see the Ironbridge described at the bottom of this page.

7. Whether you decide to explore upper Ironbridge or not, briefly describe your impressions of the town below.

   ....
   ....
   ....
   ....
   ....
   ....
   ....
   ....
   ....
   ....
   ....
   ....
   ....
   ....
   ....
   ....
   ....
   ....

8. Look at the map below and the visual evidence around you. What evidence is there to suggest that part of the town was planned and part was not?

   ....
   ....
   ....
   ....
   ....
   ....

9. List two advantages and two disadvantages of this site for a settlement in 1779.

   Advantages ....
               ....

   Disadvantages ....
                 ....

Ironbridge is all paths and passageways, a continuous up and down progression of steps and tottering, leaning walls, each turn revealing a new dark hole in a wall or a breath taking view to a horizon across the valley or three miles away. Crusty iron tie-plates and bars hold the place together, tree roots and the determined pressure of the soil pull it apart.

Extract from *Ironbridge: Landscape of Industry* by N. Cossons and H. Sowden

THE MARKET AND MARKET BUILDINGS

Documentary evidence suggests that this part of the town was developed by the Trustees of the Iron Bridge around 1784–90. Soon afterwards the Market place and Market buildings were the focal point of both the town and surrounding area.

1. Complete the record card for the Market Building on page 24.

2. What visible changes have taken place to the building since 1790?

   ....
   ....
   ....
   ....

 # THE TONTINE HOTEL

Perhaps the most immediate visual effect of the Iron Bridge on the growth of the town was the building of the Tontine Hotel at the north end of the bridge.

From the day it opened in 1784 it became the principal coaching inn in the Severn Gorge, benefiting from the increased trade that resulted from the building of the world's first Iron Bridge. Not that the owners of the hotel left trade to chance. Much was made of the Iron Bridge as a tourist attraction and the close proximity of the hotel to the bridge was stressed in advertisements placed in the newspapers of places as far away as Dublin, Chester, Bath and Bristol. The hotel, along with other businesses in the expanding town, also stressed the benefit of the new turnpike roads built to link the bridge with the local communities in and around the Gorge. In 1786 an advertisement for the Tontine stated that it had been built 'for such Noblemen, Gentlemen and others who may give preference to the new road, laid out after considerable expense, now nearly completed from Shrewsbury . . . and over the Iron Bridge.'

Look again at the extract from *Pigot's Directory* on page 20.

1. Which coach service used the Tontine as a stopping place in 1828?

   ....
   ....

2. Which other inn in Ironbridge was used as a coaching inn?

   ....

3. What evidence is there that a regular coaching service was in existence by 1828?

   ....
   ....

4. The Tontine was financed by many of the same men who financed the bridge. Do you think they were wise to put their money into two separate ventures?

   ....

RECORD CARD FOR THE MARKET BUILDING

LOCATION OF BUILDING	PRESENT USE OF BUILDING	ORIGINAL USE

DATE/PERIOD	PRESENT CONDITION OF BUILDING	DATE OF SURVEY

*c.*1790

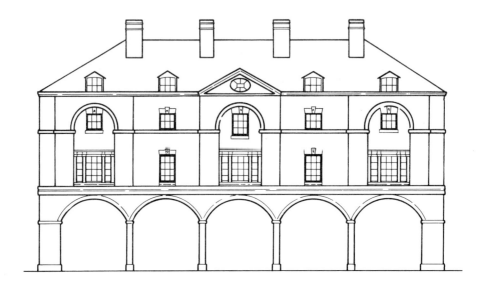

19— Sketch, photograph or postcard to be put in this box.

E   THE WHARFAGE

From the Tontine Hotel walk down Tontine Hill and along the riverside towards the Severn Warehouse. This area, between the bottom of the hill and the Severn Warehouse is known as the Wharfage. Although the buildings to be found along the Wharfage date mainly from the eighteenth and nineteenth centuries, there had been a busy settlement here long before this time consisting of cottages, inns, warehouses and riverside industries such as rope making and boat building.

1. On your walk along the Wharfage find evidence of the features listed in the key below and mark them on the map above using a different colour for each feature. Note that some buildings no longer serve their original purposes.

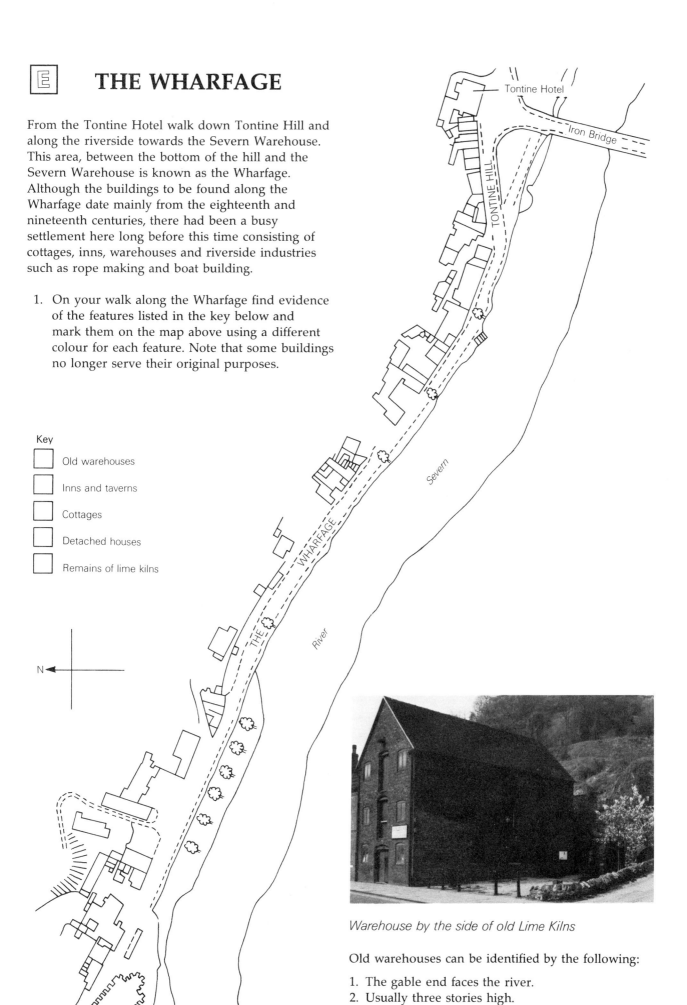

Key
- Old warehouses
- Inns and taverns
- Cottages
- Detached houses
- Remains of lime kilns

Warehouse by the side of old Lime Kilns

Old warehouses can be identified by the following:

1. The gable end faces the river.
2. Usually three stories high.
3. Some have sack hoists projecting from the top storey.
4. Close to the road.

25

SOURCE A
A view of the Wharfage by William Williams, c.1776

SOURCE B
A view of the Wharfage, c.1900

SOURCE C
A view of the Wharfage 19—. Personal photo to be put here

2. An early nineteenth-century directory listed four inns along the wharfage – The Swan, Talbot, White Hart and The Rodney. Of the four two still exist under their original names, one has changed name and the fourth no longer exists.

   Which of the inns has changed name?

   ....

   Which of the inns no longer exists?

   ....

3. Find the buildings shown in the two nineteenth-century photographs below and note two changes and two similarities between then and now.

   Changes ....
   ....
   ....
   ....

   Similarities....................................
   ....
   ....
   ....

   Changes ....
   ....
   ....
   ....

   Similarities....................................
   ....
   ....
   ....

Look carefully at Sources A and B on page 26.

4. State two similarities and two differences between the two views

   ....
   ....
   ....
   ....
   ....
   ....
   ....
   ....
   ....

5. State two similarities and two differences between Source B and the view along this part of the Wharfage today.

   ....
   ....
   ....
   ....
   ....
   ....

6. What evidence is there that the River Severn was a busy trade route at the time of Source A?

   ....
   ....

7. Is there any visible evidence today that the Wharfage no longer caters for river trade?

   ....
   ....

8. You have probably noticed that there were more changes to the Wharfage between the dates of Source A and Source B than there were between the time of Source B and today. (Note how many of the buildings you can see in B are still standing today.) Suggest three reasons for the many changes that have taken place between the dates of Sources A and B.

   (i) ....
   ....

   (ii) ....
   ....

   (iii) ....
   ....

9. Suggest three reasons for the lack of change between Sources B and the Wharfage today.

   (i) ....
   ....

   (ii) ....
   ....

   (iii) ....
   ....

10. Using the features shown in the picture below as your only source of evidence, suggest the earliest date that it could have been painted.

   ....
   ....
   ....
   ....
   ....
   ....
   ....
   ....
   ....
   ....

On your walk along the wharfage look out for a cast-iron mile post. Make a sketch of the post in the box below.

11. What is the link between the mile post, the Tontine Hotel and the Iron Bridge?

   ....
   ....
   ....
   ....
   ....
   ....

THE SEVERN WAREHOUSE

The Severn Warehouse was the point from which the Coalbrookdale company's iron goods were shipped away down river to the port of Bristol, from where they were transported to towns and villages throughout Britain and the world. Unfortunately no documentary evidence survives which allows us to date the building with certainty. But from the style of the architecture we can guess that it was built either in the late 1830s or early 1840s.

1. Find the grooves in the concrete on the wharf by the side of the Warehouse. What do you think was the purpose of these grooves?

   ....
   ....

Today the Severn Warehouse is part of the Ironbridge Gorge Museum and displays a short audio-visual programme on the history of the Gorge and the various museum sites. There is also a display on river transport and a cross section of a middle-class home of 1900 illustrating the effect of the industrial revolution on everyday objects around the house. The entrance lobby also serves as a bookshop where postcards and museum guides can be purchased, among other things.

2. Below is a sketch of about 1835 showing the river bank opposite where the Severn Warehouse now stands. Although very amateurish, this sketch is a valuable source of evidence about a wide variety of subjects. Look carefully at the picture and identify the features labelled around the edge. Draw arrows to these features.

A late nineteenth-century photograph of the Severn Warehouse and a trow waiting by the wharf.

3. What now remains of the many features shown in this picture?

   ....
   ....
   ....
   ....
   ....
   ....
   ....
   ....
   ....

Horse gin for raising people from a local mine   *Boat building yard*

The Iron Bridge   *Loading crane*   *Severn trow*

CONCLUSIONS AND FURTHER RESEARCH

It is hoped that the fieldwork exercise you have just completed has helped with your understanding of how and why a community developed along the banks of this part of the River Severn. In particular you should be aware of the value of physical remains as evidence for studying the past. From these we can find out much about the past providing we observe and interpret them carefully. Like all sources of information and evidence, however, physical remains have their limitations. For example, the Iron Bridge provides valuable information about eighteenth-century bridgebuilding techniques, but, no matter how much we investigate the physical remains, we will not find the answers to questions such as why was it built and how much did it cost? For answers to these and many other questions about the bridge and the local community we need to examine other types of sources.

POINTS FOR DISCUSSION

1. Tick the appropriate box.

How useful is the Iron Bridge as evidence of:	Very useful	Useful	Of little use
(i) Eighteenth-century bridgebuilding techniques;			
(ii) Abraham Darby III's involvement in the project;			
(iii) how the bridge was constructed;			
(iv) the reasons for building the bridge.			

2. What do you think are the advantages and limitations of using physical remains as your ONLY source of information about the history of the area you have just explored.

3. A historian trying to reconstruct a picture of life in the Severn Gorge during the eighteenth and nineteenth centuries would use a variety of sources. Suggest three types of sources, in addition to physical remains, that could be used and for each one explain how it might be of use.

RESEARCH WORK INSIDE THE SEVERN WAREHOUSE MUSEUM

The aim of your visit to the museum is to discover more about the Severn Gorge in the late eighteenth and early nineteenth centuries. There are two small display areas and a short audio-visual programme. After listening to this programme complete the following tasks:

1. Find the display area about the River Severn and make notes about

    (i) the type of vessels used on the river during the eighteenth and nineteenth centuries;

    (ii) the importance of the river to the development of the Severn Gorge;

    (iii) people associated with river trade such as the boat builders.

2. By the early nineteenth century the Severn Gorge was being referred to as 'the most extraordinary district in the world'. Examine the various artistic and written accounts of the area and write down the impression they give of the area during the period. What is your impression of the Gorge today?

THE CONSERVATION AND PRESERVATION OF THE IRON BRIDGE

In addition to studying physical remains to find out about the past the historian is also interested in conserving and preserving such remains for the future. This is important so that future generations may have the chance to make their own interpretations of the past. Many people also gain great pleasure and enjoyment from visiting these sites.

The preservation of industrial remains only arises in certain cases where a monument is judged to be of outstanding historical value. But even then preservation is not automatic. There are many factors to be considered, in particular time and money. The Ironbridge Gorge Museum Trust, which is a private charitable trust, has had to raise thousands of pounds to preserve the many industrial monuments in and around the Severn Gorge. Perhaps you have noticed that there has been considerable conservation work on the Iron Bridge itself. The problem is the unstable geological conditions of the Gorge. The bridge is built on shale and mudstone. Earth slips occur frequently, and since 1779 these earth movements have weakened the structure of the bridge causing fracturing and distortion of the cast iron. At one point in the 1960s this had become so bad that it was decided that the only permanent solution to the problem was to stop the abutments moving forward towards the river.

In 1972 a framework of concrete and steel was erected inside the north abutment of the bridge and then in 1973 an inverted concrete arch was constructed in the river bed to keep the abutments apart. By 1974 the work was finished at a cost of over £50,000 and the bridge was safe.

1. What evidence can you see today of the restoration work?

2. Make a list of arguments for and against the government spending money on conserving and preserving historical building and sites?

3. What are your opinions and feelings about conserving historical buildings?

The Iron Bridge during restoration

ADDITIONAL TASKS IN THE CLASSROOM

1. Find out more about the Industrial Revolution and, in particular, about other settlements that developed during the period 1700–1850. Compare and contrast these with the town of Ironbridge.

2. Visit your local library or record office to find a copy of a local trade directory of your area in the early nineteenth century. Use the list of occupations to reconstruct a picture of your area in the past. Follow this up by using other sources such as maps and newspapers, not forgetting to undertake an exploration of the physical remains.

3. Find out about bridge construction past and present and compare the advantages and disadvantages of the various designs of bridge with the one used for the Iron Bridge.

4. Find out why the town of Ironbridge is no longer the prosperous community it was in the early nineteenth century.

FURTHER READING AND ACKNOWLEDGEMENTS

N. Cossons and B. Trinder, *The Iron Bridge*, Moonraker Press and Ironbridge Gorge Museum Trust, 1979

N. Cossons and H. Sowden, *Ironbridge: Landscape of Industry*, Cassell, 1977

W. Grant Muter, *The Buildings of an Industrial Community: Coalbrookdale and Ironbridge*, Phillimore, 1979

L. Metcalfe, *Discovering Bridges*, Shire Publications, 1970

B. Trinder, *The Most Extraordinary District in the World: Ironbridge and Coalbrookdale*, Phillimore, 1979

3-D Cut-out Model
The Iron Bridge 1781, Landscape Models, Salop, 1979

Teachers Pack
The Iron Bridge, Ironbridge Gorge Museum Trust, 1980

Slide Tape Programme
Portrait of a River: River Severn (RGB/458), Student Recordings Ltd, 88 Queen St, Newton Abbot, Devon

Much useful background information for this book came from the original research of Barrie Trinder and Neil Cossons.

The author and publishers would like to thank the Ironbridge Gorge Museum for illustrations from their collection and permission to reproduce them. The illustration on page 28 is by courtesy of Shrewsbury Museums Service.

Text © G. A. Alton 1988
Illustrations © Stanley Thornes (Publishers) Ltd 1988

All rights reserved. No part of this publication may be reproduced, stored in a retrieval system or transmitted in any form or by any means, electronic, mechanical, photocopying, recording or otherwise, without the prior written consent of the copyright holders. Applications for such permission should be addressed to the publishers: Stanley Thornes (Publishers) Ltd, Old Station Drive, Leckhampton, CHELTENHAM GL53 0DN, England

First published in 1988 by:
Stanley Thornes (Publishers) Ltd
Old Station Drive
Leckhampton
CHELTENHAM GL53 0DN
England

Typeset by Tech-Set, Gateshead, Tyne & Wear
in 10/12 Palatino
Printed and bound in Great Britain by
Ebenezer Baylis & Son, Worcester

British Library Cataloguing in Publication Data

Alton, G.
 The Iron Bridge.
 1. Shropshire. Ironbridge. Iron. Iron Bridge
 I. Title
 624'.67'0942456

ISBN 0-85950-692-4